Leaping Learners
Education, LLC

Follow us on Facebook
@leaping.learner

Please leave us a review on Amazon if you enjoy the book!

Published by Leaping Learners Education, LLC, Verona, NJ 07044 .

Every attempt has been made to credit each photo. Please contact us if there has been an error and we will resolve the issue.

Photo Credits
Owl graphics © *Skop*/shutterstock.com

Cover, © *ArCaLu*/shutterstock.com; title © *Dennis W Donohue; PG.2* © *ArCaLu*/shutterstock.com; PG 4 © max_776/stock.adobe.com; PG. 6 © *ArCaLu*/shutterstock.com; *PG. 8* © *Lars Ove Jonsson*/shutterstock.com; *PG. 8* © *Lars Ove Jonsson*/shutterstock.com; *PG. 10* © *Shaber*/shutterstock.com; *PG. 12* © *Volodymyr Burdiak*/shutterstock.com; *PG. 14* © *critterbiz*/shutterstock.com; *PG. 16* © *Tomas Palsovic*/shutterstock.com; *PG. 18* © *AndreAnita*/shutterstock.com; *PG. 20* © *Martin Rudlof Photography*/shutterstock.com; *PG. 22* © *DPS*/shutterstock.com; *PG. 24* © *Gleb Tarro*/shutterstock.com; *PG. 26* © *Vaclav Zilvar/adobestock.com; PG. 28* © *ArCaLu*/shutterstock.com; *PG. 30* © *FuFioCat*/shutterstock.com

All designs by Sean Bulger
All other pictures by Sean Bulger or royalty-free from Pixabay.com

ISBN:
978-1-948569-33-0

Dear Parents and Guardians,

Thank you for purchasing a *Ozzy Learns About* series book! After teaching students from kindergarten to second grade for more than seven years, I became frustrated by the lack of engaging books my students could read independently. To help my students engage with nonfiction topics, my wife and I decided to write nonfiction books for children. We hope to inspire young children to learn about the natural world.

Here at Leaping Learners, LLC, we have three main goals:

1. Spark young readers' curiosity about the natural world
2. Develop critical independent reading skills at an early age
3. Develop reading comprehension skills before and after reading

We hope your child enjoys learning with Ozzy. If you or your children are interested in a topic we have not written about yet, send us an email with your topic, and maybe your book will be next!

Thank you,

Sean Bulger, Ed.M

Reading Suggestions:

Before reading this book, encourage your children to do a "picture walk," where they skim through the book and look at the pictures to help them think about what they already know about the topic. Thinking about what they already know helps children get excited about learning more facts and begin reading with some confidence.

Preview any new vocabulary words with your child. Key vocabulary words are found on the last few pages of the book. Have your children use each new phrase in their own words to see if they understand the definition.

After previewing the book, encourage your child to read the book independently more than once. After they have read it, ask them specific questions related to the information in the book. Encourage them to go back and reread the relevant section in the book to retrieve the answer in case they forgot the facts.

Finally, see if your child can complete the reading comprehension exercises at the end of the book to strengthen their understanding of the topic!

This book is best for ages 6-8
but. . .
Please be mindful that reading levels are a guide and vary depending on a child's skills and needs.

Ozzy Learns About . . . Grizzly Bears

Written by Sean Bulger

Table of Contents

Hi! My name is Ozzy.
I love to discover and
learn new things.
In this book, we will
learn about grizzly
bears. Let's go!

Introduction

Watch out! It's a grizzly bear!

Grizzly bears look cuddly, but they are very strong and dangerous animals.

Habitat

Grizzly bears can only be found in parts of North America.

ARCTIC OCEAN
EUROPE
ASIA
AFRICA
PACIFIC OCEAN
AUSTRALIA
ATLANTIC OCEAN

Grizzly bears live in forests among the trees and bushes.

Grizzly bears make **dens** in the forest.

They make their dens by digging holes in the ground or piling leaves near fallen trees.

Den

Grizzly bears **hibernate** in the winter.

This means they sleep safely in their dens during the winter and come out in the spring.

Tricky word:
Hibernate
say...
Hi-ber-nate

Body

Grizzly bears are mammals.

They are covered with fur and have four legs.

Grizzly bears have large, sharp teeth.

They use their teeth to protect themselves and to eat their food.

Teeth

Grizzly bears have long, sharp claws.

They use their claws to hunt for food, climb trees, and protect themselves.

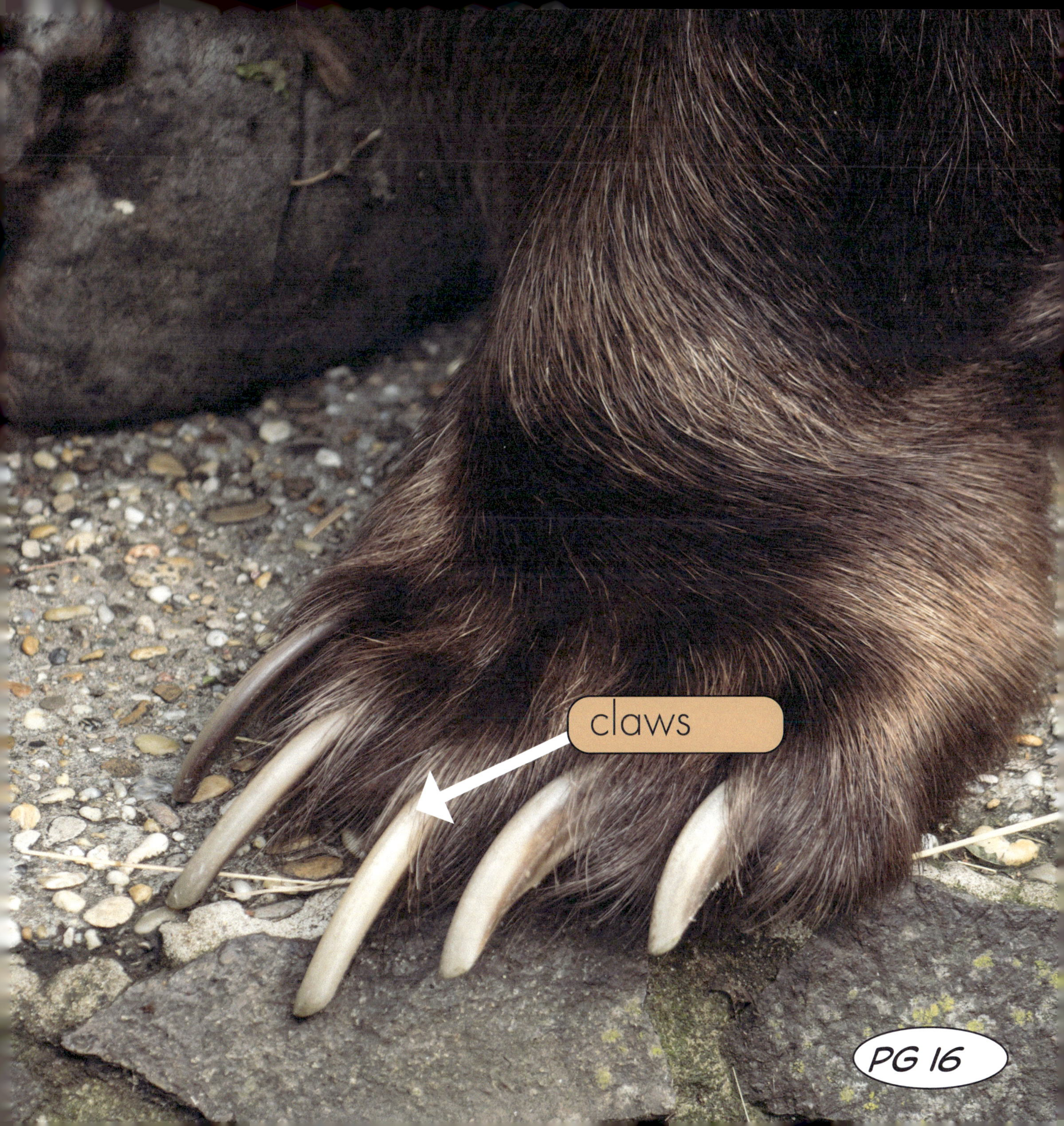
claws

Grizzly bears are fast runners and swimmers. They use their speed to help them hunt for food.

Diet

Grizzly bears are **omnivores**.

This means they eat meat, berries, and plants.

Salmon are type of fish grizzly bears love to eat.

Salmon

Grizzly bears hunt for salmon in nearby rivers. They catch the salmon as they swim upstream.

Grizzly bears also eat mammals such as deer and elk.

PG 26

Babies

Grizzly bears give birth to one or two babies a year called **cubs**.

A mother grizzly bear will protect her cubs from any threat.

PG 28

Under Threat

Grizzly bears are threatened because people hunt them and are cutting down the forests they live in.

PG 30

National parks help to keep grizzly bears safe.

Yellowstone National Park is one park where grizzly bears are safe from **deforestation** and hunting.

YELLOWSTONE
NATIONAL
PARK
NATIONAL
PARK
SERVICE

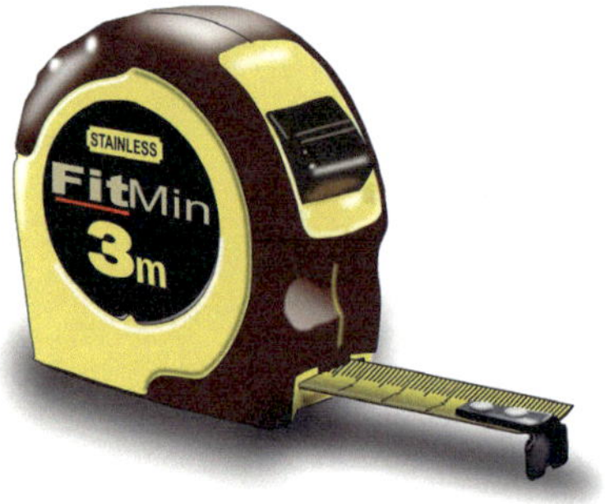

A grizzle bear can be up to 8ft tall when it stands on it's back legs

Grizzly bears can run up to 30 miles per hour!

Grizzly bears eat up to 90 pounds of food a day. That is like eating 360 quarter-pound burgers a day!

Glossary

A glossary tells the reader the meaning of important words.

Den – A grizzly bear's home made underground or by piling up twigs and leaves

Hibernate – When an animal puts it's body in a deep sleep for the winter

Omnivores – Animal that eats plants and meat

Cubs – Baby bears

Deforestation – The systematic cutting down of trees over a large area

Draw a picture of a grizzly bear

Draw a picture of a grizzly bear eating something

The grizzly is eating ____________________.

Quiz

1. Which continent do grizzly bears live on?
a. South America
b. North America
c. Asia

2. What is a grizzly bear's home called?
a. Den
b. Burrow
c. Nest

3. How do grizzly bears hunt for salmon?
a. From the shoreline
b. When they swim upstream
c. By using traps

Common core standards:
RI. 1. 1 - Questions 1, 2
RI. 1. 2 - Question 3

4. How many cubs does a grizzly bear have at a time?

a. 8–0

b. 11–24

c. 1-2

5. True or false: Grizzly bears can climb trees?.

a. True

b. False

6. What does the picture on page 10 show?

a. How grizzly bears hibernate

b. What a grizzly bear's teeth look like

c. What grizzly bears eat

Common core standards:
RI. 1. 1 - Questions 4, 5
RI. 1.8 - Question 6

Want to learn about rainforest animals? Check out the "Matt Learns About..." series!

Great for emerging readers ages 6–8